Illumen
Winter 2021

In this issue:

Article by William Landis: Richard Wright's Haiku

Interview with Alan Ira Gordon

Featured Poet: Avra Margariti

The Meg Smith Page

The Guy Belleranti Page

Illumen
Winter 2021

Edited by Tyree Campbell

Cover art "The Embrace" by Denny Marshall
Cover design by Laura Givens

Vol. XVII, No. 2					January 2021
Illumen [ISSN: 1558-9714] is published quarterly on the 1st days of January, April, July, and October in the United States of America by Hiraeth Publishing, P.O. Box 141, Colo, Iowa, 50056-0141. Copyright 2020 by Hiraeth Publishing. All rights revert to authors and artists upon publication except as noted in selected individual contracts. Nothing may be reproduced in whole or in part without written permission from the authors and artists. Any similarity between places and persons mentioned in the fiction or semi-fiction and real places or persons living or dead is coincidental. Writers and artists guidelines are available online at www.albanlake.com/guidelines. Guidelines are also available upon request from Hiraeth Publishing, P.O. Box 1248, Tularosa, NM 88352, if request is accompanied by a SASE #10 envelope with a 55-cent US stamp. Editor: Tyree Campbell. Subscriptions: $28 for one year [4 issues], $54 for two years [8 issues]. Single copies $8.00 postage paid in the United States. Subscriptions to Canada: $22 for one year, $40 for two years. Single copies $10.00 postage paid to Canada. U.S. and Canadian subscribers remit in U.S. funds. All other countries inquire about rates.

New from Terrie Leigh Relf!!
Postcards From Space

Terrie Leigh Relf loves sending and receiving postcards from the four corners of the universe—and beyond! Postcards tell a story. They are mementos from friends and family—and from total strangers—and provide a glimpse into life's journeys, observations, and adventures.

Here are some messages on postcards from space, found aboard a derelict craft that crashed on an arid, lifeless world. The OSPS (Outer Space Postal Service) has delivered these messages to Terrie, who now presents them to you. This is what it is like out there.

https://www.hiraethsffh.com/product-page/postcards-from-space-by-terrie-leigh-relf

A Little Help, Please

In the world of the small indie press we fight a never-ending battle for attention to our work, as writers and in publishing. Here's an example: big publishers [you know who they are] have gobs of $$$ that they can devote to advertising and marketing. Here at Hiraeth Publishing, our advertising budget consists of the deposits for whatever soda bottles and aluminum cans we can find alongside the highways. Anti-littering laws make our task even more difficult . . . ☺

That's where YOU come in. YOU are our best promoter. YOU are the one who can tell others about us. Just send 'em to our website, tell them about our store. That's all. Just that.

Of course, we don't mind if you talk us up. We're pretty good, you know. We have some award-winning and award-nominated writers and artists, plus other voices well-deserving to be heard [not everyone wins awards, right?] but our publications are read-worthy nevertheless.

That number once again is:

www.hiraethsffh.com

Friend us on Facebook at Alban Lake's Hiraeth Page and follow us on Twitter at

@albanlake and @albanlakepub

Contents

Features

11	The Meg Smith Page
13	Alan Ira Gordon interviewed by Terrie Leigh Relf
26	Article: Nightmares of a Native Son: The horrorku of Richard Wright by William Landis
38	The Guy Belleranti Page
39	Featured Poet: Avra Margariti
55	Who's Who

Poems

7	Intent by Pamela Love
8	All Our Faults Are Fallen Leaves by Eric Robert Nolan
12	Damsels of the Next Apocalypse by Russell Hemmell
22	The Gallant Moon by Stephanie Smith
23	Smile by Lori R. Lopez
29	In the Time of Golden Trees by Krista Canterbury Adams
30	Offering by Colleen Anderson
	haiku by Debby Feo
31	The Starfarer by K. R. Lehman
35	Am I made the same as you? by Katherine Relf-Canas
37	Not of Byzantium by Eric Robert Nolan
46	Rumpled by Baishampayan Seal
47	Zombies by Sean Ferrier-Watson

48	The Color Green by Stephanie Smith
49	In Feline Grace by Colleen Anderson
50	Delivery by Lori R. Lopez
53	Finally by Krista Canterbury Adams
54	Graveyard Waltz by Stephanie Smith

Illustrations

45	Untitled by Baishampayan Seal

Intent
Pamela Love

I knew a girl who raged at stars
And hurled insults upon the Moon.
By day and night she cursed the Oort
And slandered all of Saturn's rings.
No galaxy outranged her wrath—
No comet passed but felt her scorn—
No longer did they feel their orbits Home.
Astronomers could only weep...
Her death was swift, and no surprise.
She saw the meteor approach.
(Her final words were well-rehearsed.)
She spoke her last: "I made you look."

All Our Faults Are Fallen Leaves
Eric Robert Nolan

Again an annual angled auburn hand
announces advancing Autumn —
fingers aflame, the first Fallen leaf,
As slow in its descent, and as red,
as flailing Lucifer.

Hell in our sylvan vision
begins with a single spark.
The sting of the prior winter
subsided in July,
eroded at August.
Now, as at every September,
let new and cooler winds
fan a temperate flame.

May this nascent season only
bring brick-tinted perdition
and carmine Abaddon.
Where flames should burn, may there be
only rose tones on wide wine canvasses,
tormentless florid scarlets,
griefs eased in garnet trees.

What I hold in my heart to be true
is Edict at every Autumn:
Magentas may not make
forgetful a distracted God,
unless we ourselves forget
or burn to overlook.

Auden told us "One Evening"
to "Stand, stand at the window,"
and that we would love our neighbor,
but he didn't counsel at all
about how we should smolder there.

Outside my window, and yours,
if the Conflagration itself
acquits us all by claiming only
the trees upon the hill,
the Commonwealth a hearth,
Virginia an Inferno,

Then you and I
should burn in our hearts to absolve
ourselves and one another,
standing before the glass,
our curtains catching,
our beds combusting,
our bureaus each a pyre.
Take my hand, my friend, and smile,
there on the scorching floor,
beneath the searing ceiling and
beside the blackening mirror
that troubles us no longer,
for, about it, Auden was wrong.
God's wrathful eye
will find you and I
incandescent. The damned
are yet consigned to kindness.
All our faults are Fallen leaves.
Forgive where God will not.

Out of our purgatory
of injury's daily indifference,
let our Lake of Fire

be but blush squadrons of oaks,
cerise seas of cedar, fed
running ruby by sycamore rivers,
their shores reassured
by calm copper sequoias,
all their banks ablaze
in yellowing eucalyptus.

Let the demons we hold
harden into bark
holding up Inferno.
All their hands are branches now;
all their palms are burning.

There, then, softly burning, you and I,
may our Autumn find
judgmentless russets,
vermilion for our sins,
dahlia forgiveness,
a red for every error,
every man a love,
every love infernal,
and friends where devils would reign.

— Author's note: the poem to which I've responded above, with its images of standing at the window and the mirror, is W. H. Auden's "As I Walked Out One Evening."

The Meg Smith Page

Falling Flowers

A woman stands
in a field of heather.
She wears her cuts
inside and out,
in the music of cuts
without bleeding.
We, all of us, are more than
highways of blood
or where someone
has tried to open us.
We are no fantasy
of petals, but are grown from blood
of our own true sun.

The Hidden Altar

We drew the circle of history
in our five points.
I did not know it would
capture time.
I did not know it would enfold
our brokenness.
I still call to something
deeper than grass --
with dragons like
beacons in the far fog.

Damsels of the Next Apocalypse
Russell Hemmell

Eyes of fairies, from mayhem and fight
eyes of villains, devoid of grace
hardened hearts, skies out of sight
we are Death, showing their face.

Justice is ours, we faithless sprites
escaped from the woods in a faraway land
taking with us a mirror and a scythe
knowing since the onset how the story will end.

Four bloodstained letters and we unleash doom
four vivid colours and a palette of spells
four clear voices where the darkness does loom

for finding ourselves all sitting down here
at the round table of the green-haired ghouls
of a world that forgot the value of fear.

A Day in the Life Interview Series: Alan Ira Gordon
with Terrie Leigh Relf

What types - and forms - of writing do you do? If you're also an editor, what is your niche?

I mostly write science fiction and fantasy poems and short stories, with an occasional light horror or mainstream story or poem in the mix. And I write some historical articles and occasional essays or columns. I've done some guest editing in the genre, both Issue #24 of Eye To The Telescope, the on-line electronic quarterly journal of the Science Fiction Poetry Association (SFPA) and Strange Summer Fun, a short story anthology from Whortleberry press.

What is your area(s) of subject matter expertise? How did you discover this niche? What intrigues you about it?

Professionally, I'm an urban planner and urban planning instructor at Worcester State University, but I've been writing for a lifetime. For a long time, I just wrote short stories, but about a decade ago I realized that so many of my ideas worked better as poems rather than stories and my creativity was better channeled and took off from there.

I'm very intrigued by the challenge of writing within the structure and brevity of a poem while getting across to the reader a theme, storyline and point of view. It's like doing a puzzle every time and I love it!

How do you balance your creative and work time?

I'm very fortunate to be semi-retired at this stage of my career, teaching part-time. That frees me up for plenty of time for writing efforts.

What tips do you have for other writers and/or editors?

One tip was given to me a long time ago by the great science fiction writer Jack Dann when I met him at one of the annual ReaderCon conventions. He told me to consider going back to draft stories even years after you've put them aside and giving it another go. Even though he was talking about story-writing, I think it applies to any form of writing (poems, etc.). He said that your new, fresh perspective or time passed will help finish the tale and sure enough, he's right-it's worked for me well. A second tip is to not be afraid to re-draft and re-write. So many writers I've met feel that an initial draft or two is always the finished product and that's not so. I know a very literate, accomplished mainstream poet who had a wonderful poem recently published in a very prestigious journal and she had revisited and redrafted her poem over a two-year period. It takes a confidence and patience that many writers aren't comfortable with, but it's so worth it at times.

What are your thoughts on the creative process in general and your creative process in particular?

I think that every person's creative process is different and unique to their own personality, style and experiences. While folks can get lots of "how to write" or "how to be creative" advice, take it all with a grain of salt and just do what feels right for you. My own process keys-in on being patient, taking my time as I think about a plotline, start drafting and keep re-drafting, and not rushing to a premature wrap-up.

Where do your ideas come from?

It's funny, they come from everywhere! I've had poem and story ideas pop into my head from conversations with people, from reading and watching tv, seeing or reading the news, etc. This past year, I've actually gotten good ideas for poems just from reading an on-line "word of the day" website that I like to check-out. Steve Martin says it best when he says that "you are an idea machine, with a million thoughts a day each representing a story idea."

Where have you been published? Upcoming publications? Awards and other accolades?

My short stories have been published more in anthology collections than magazines, quite a few over the years in collections from Alban Lake Publishing, Whortleberry Press and Hiraeth Publishing, to name a few. My poems have a much wider publication range, from the more known genre magazines such as Analog, The Magazine of Fantasy & Science Fiction and Star*Line to some wonderful smaller publications including Illumen, Disturbed Digest, Pedestal Magazine and Outposts of Beyond. I enjoy it when my writing is published in expected publications such as Dog Eyes Magazine and Worcester Magazine.

Award-wise, my poetry has received five SFPA annual Rhysling Award nominations, one Dwarf Star nomination and won an Analog year's award (second prize). Last year, my science fiction/fantasy poetry collection "Planet Hunter" received an annual Elgin Award nomination. One of my mainstream fiction stories won the annual Worcester Magazine story competition and another story won the annual Whortleberry Press year's best science fiction award. Many years ago, I was humbled that the late great editor/writer Gardner Dozois gave one of my stories an Honorable Mentions listing in his eighth annual Year's Best Science Fiction story collection (St. Martin's Press).

What are you working on now?

I always have a few different drafts in the mix. Right now, that includes a fantasy short story, a few science fiction poems and a mainstream essay. I'm also writing a book review of a new urban planning textbook for a publisher to use in the text's on-line marketing.

What challenges have you faced as a writer and/or with a particular project? How did you meet them? What did you learn from these challenges and how did they make you a better writer and/or editor?

Two challenges immediately come to mind. When I was writing a gothic horror story ("Vincent and Paul In The Yellow House" published in The Night Café anthology edited by Tyree Campbell, Alban Lake

Publishing), I needed to be as historically accurate as possible within my fictional setting, regarding the lives and times of Vincent Van Gogh and Paul Gauguin. I learned to be patient and research, research and research. I spent about two years researching prior to going beyond a very preliminary story draft and it was well-worth it.

The second challenge was as an editor when I guested-edited a few years ago Issue #24 of Eye To The Telescope, the on-line quarterly poetry journal of the Science Fiction Poetry Association (SFPA). The challenge was to go beyond selecting poems for that issue based on what I personally enjoyed, but to focus rather on a wide variety of poetry selected as well as choosing based upon what I believed reader's would want to have the opportunity to read. Focus on those criteria I would hope made me a more responsible editor.

Are you plotter and planner or a discovery writer?

Actually, I've been each at different times and with different projects, I think it depends on the nature of the plot or story/poem/essay idea.

Are you currently a writing mentor? If so, what are your thoughts on mentoring?

No, I've been mentored myself at times but never had the experience of being the mentor.

Are you currently, or have you ever, been in a writing group? Your thoughts?

I'm in a wonderful monthly poetry group at Worcester State University which includes faculty, students and folks from across the community. It's on hiatus now with the pandemic but I hope it comes-back eventually. My thought is that it's wonderful to be exposed in such a group to such a diverse array of people, points of view and types of poetry, I cherish every moment in the group and highly recommend the experience!

I know our readers would love to hear about your networking, marketing, and promotional experiences - including tips.

I've gotten great writing and marketing advice from conversations with writers and editors who I've met over the years at various ReaderCons. Such folks who've given me worthwhile advice and direction over the years include Jack Dann, Glen Cook, esteemed editor/writer Mike Resnick, F&SF publisher Gordon Van Gelder and Robert Sawyer, to name a few. My advice is don't hesitate to ask for such tips from such experienced writers and editors. Don't be intimidated, these folks are approachable and are very willing to talk shop at these writing and/or convention events.

Your thoughts on having an agent?

I've never done so. I think it's only worthwhile if you're trying to get a novel published with a major publisher.

Your thoughts on self-publishing?

I don't think much of it. I've read that the average self-published novel or anthology is lucky to sell about 40 copies and those are mainly to the author's family and friends. The only time it seems worthwhile is when the author is a very established best-selling professional who is balancing their publications with a few self-published efforts to bypass their usual publisher middleman.

Anything else you'd like to add that I haven't asked? For example, what would you like to see more of in your specific genre? In the publishing field? Where do you see yourself in the next year? Next five years? etc.

Just one piece of advice to add: trust in and work with your editors. I'm been blessed with some great editors over the years who not only have published my writing but when appropriate have worked with me with revision/re-write suggestions and ideas. It's not easy to accept those critiques of our work but in the long run it's made my products better and much appreciated. So a quick thanks to some of those past and present editors including Tyree Campbell, Gordon Van Gelder, Marge Simon, FJ Bergman, Vince Goleta, Emily Hockaday, Jean Goldstrom and the late great Gardner Dozois.

Your questions have been far-ranging and excellent for both writers and readers. I guess in the next year, 5 years, etc., I see myself happily doing exactly what I'm doing right now, reading, thinking about ideas and writing, writing, writing (and occasionally guest-editing)!

Planet Hunter
By Alan Ira Gordon

Alan Ira Gordon is an urban planner and urban studies professor at Worcester State University and writer of science fiction/fantasy short stories and poetry. He's a three-time Rhysling Award nominee and a Dwarf Star Award nominee. He's contributed to several Alban Lake publications, is a frequent contributor to Star*Line and guest-edited Issue #24 of Eye To The Telescope, the on-line publication of the Science Fiction & Fantasy Poetry Association (SFWA). His poetry, short stories and articles have been published in various genre magazines and anthologies, a partial list of which can be found on his webpage at www.alaniragordon.com.

https://www.hiraethsffh.com/product-page/planet-hunters-by-alan-ira-gordon

The Gallant Moon
Stephanie Smith

I prayed all day for the long, dark night
to come and rescue me
as daylight held me by the ankles,
drowning me in broken promises,
the temperature boiling hot

I prayed the moon could comfort me
Rise high and mighty, silver sword in hand,
atop His majestic steed
And slay the sun that has gone on too long
blinding and burning me

I embraced the moon when He came into view,
let Him drape me in His velvet cape
He parted the sea of stars to clear
a path for us to sail

Away, away across the universe
I left it all behind
Around the world and far away
as the planets were pushed aside

The Earth fell into spasms
and lost its gravity
I waved goodbye to the ocean tides
Forever free, I set my sights
on fresh, new galaxies

Smile
Lori R. Lopez

A black and white day
Like an old photograph that has
History, once a scrap of memory, the record
Of an event or simple candid moment
Trapped behind glass
A conspicuous fish to be gazed at in a bowl
On a table or shelf. There's a tree
Near the edge of the frame
Unable to avoid being included, part of the
Scene decorating a mantel

Tucked in a musty album or scrapbook

I see a weather-beaten facade
Unpainted, filling the background
Two stark figures at the fore, side by side
Nowhere to escape, squinting curiously
Toward the lens of an outsider. Clad in faded
Tatterdemalion threads; ragamuffin country
urchins
Apple-cheeked and guileless. A little shy
Or is it fear? They couldn't know their aspects
Would be captured, collected as
Souvenirs. *Pose for me.*

A glib request; a disarming grin

The stranger tall and elegant
As men and women appear in colorless classic

Films; the kind who aren't roughhewn,
Stoic, coarse-mannered. He snaps a photo
With nonchalance, an easy casual air, and tips
A brim, a gray Fedora
Dark against a white sky. The day
Has a neutral overcast feel. Turning
He climbs in his automobile, aristocratic
Self-assured, smug of demeanor

Wearing a suit and a smirk

The boy and girl wave, staring while he
Motors away down an empty stream of asphalt
Then shrinks to a dot, a mere glint that
Flares and is gone. The kids
Do not resume playing, chasing in a circle
Kicking dust — but stand immobile —
Statues. They cannot forget the Photographer
Who pulled off the road on a whim and
Will not forget them, for his camera
Took more than their picture

It snatched who they are

Faceless, they no longer resemble
Brother and sister but a pair of featureless
Store Mannequins, robbed of
Matching eyes, intrinsic looks, a key factor
In their identities if not their humanity
What made them special. Removed, lost in the
Theft, erased. Twin mugs having shaped how
Each viewed the other and themselves
Extracted simultaneously through a shutter
A mechanical Photo-Synthesis

I watched it all behind glass

My mouth agape. A witness, seated
On the passenger-seat of a Family Sedan
Parked outside The Sunup Cafe where
My mother had stopped to order food
We would eat in the car to avoid paying a tip
I thought the Waitress in the window
Deserved better and sympathized before
Noticing the man arrive
His License Plate read SMILE
The guy winked at me, striding to depart . . .

Young, less certain, I said and did nothing

Until I found the portrait on a wall
By then it was far too late.

Nightmares of a Native Son: The horrorku of Richard Wright
By William Landis

The artist must bow to the monster of his own imagination. –Richard Wright

When you think of Richard Wright, author of the classic novel *"Native Son"* and his memoir *"Black Boy"* you think of his controversial, and unsettling depiction of the African American experience in the early 20th century. His name is mentioned among the pantheon of the Harlem Renaissance, and as a legend among American authors, but little is mentioned about his ventures in the Haiku form or more precisely, being one of horrorku`s forefathers.

Richard Wright was born on September 4, 1909 on a plantation in Roxie, Mississippi. Despite coming from a broken home he managed to excel academically, and sell his first story at the age of 15, *"The Voodoo of Hell`s Half-Acre"* to the *Southern Register*, a local black newspaper. Wright followed the trends of the Great Migration, and found himself in Chicago, where he established himself in the communist literary scene. In 1946 he moved to Paris, France, where he died November 28, 1960.

In his last years of life Wright became well acquainted with the Haiku form writing more than 4,000 of the poems. 817 of those poems are anthologized in *"Haiku: The last poems of an American Icon"* by Richard Wright. The poems have diverse subjects. Below are a few that have a horror theme:

140
A spring pond as calm
As the lips of the dead girl
Under its water

145
A bright glowing moon
Pouring out its radiance
Upon tall tombstones

150
Late one winter night
I saw a skinny scarecrow
Gobbling slabs of meat

152
After seven days,
The corpse in the coffin
Turned on its side

467
A radiant moon
Shining on flood refugees
Crowded on a hill

    What Wright wrote is what we might today consider as horrorku. It would be a huge stretch to assume he was the first to write one, but fair to call him one of the forefathers of the form. He certainly took many liberties of imagination with the form, which by tradition is set in a moment of reality. So why would a writer, whose a majority of his works were based on the Black experience with racism, and poverty in America, experiment with horror themed Japanese minimalist poetry form?

Richard Wright's works were obsessed about race by necessity of his reality in the country of his birth. His move to Paris changed that. Wright once stated he "felt more freedom in one square block of Paris than there is in the entire United States of America". One could assume that living in a city where his race wasn't a burden allowed him to see the world beyond the lens of blackness and consider what horror was beyond his race.

Special thanks to **Juel Duke** for introducing me to the horrorku of **Richard Wright**

The Miseducation of the Androids
By William Landis

What happens when androids confront concepts inconsistent with their programming? William Landis examines this question by means of flash fiction and haiku that you will find pithy, poignant, and amusing.

William Landis is a science fiction poet from North Carolina. He is a graduate of North Carolina A&T State University, completing both undergraduate, and graduate work in agriculture. He is currently working on a vermicomposting project, and he is an Army reserve engineer officer. He enjoys running, writing, reading, and exploring new places.

Print: https://www.hiraethsffh.com/product-page/miseducation-of-the-androids-by-william-landis

In the Time of Golden Trees
Krista Canterbury Adams

We imagine to hunt & forage
on the sandy streambank,
to follow the paved path
& burst out through the trees.
But the Fox Mother
has shown us her teeth,
taught us to be,
to peel away
with claw & fang.

We are not the same
savages we once were
in the waving honey-leaves.
Tonight,
in this bronze wood
we are silent & broken,
our time of golden trees
has passed.

Offering
Colleen Anderson

dark pool
mirror reflecting
no stars but depth beyond
claret, pinot, cabernet
god's nectar from a hint
of sour grapes to the sweetest
smoky aftertaste
too few sips and the elusive fairy
evades too many
the spirit brings on possession
tang on tongue a warning
treat this smooth liquid
colored garnet, ruby
as the heart's desire
blood fills the night

Visions change
And dreams die
Sometimes
I hold on too long

    Debby Feo

The Starfarer
K. R. Lehman

We set out across the Milky Way
From the famed port at Mars
Solar wind filling our sails
And navigating by the stars

That serve as map and compass
Stepping stones of light
Waypoints and signal fires
Leading through the night

Pushed along by astral currents
Riding electromagnetic waves
Our destination seems no closer
But one fact from madness saves

A fact I oft remind the crew
That this journey will take years
But oh will it be worth it
For we are pioneers

The first to sail the cosmos
And brave the empty space
That imposingly surrounds the sphere
Of our esteemed birthplace

And bolstered now their spirits rise
And onward we sail through night
But in my captain's quarters
I grimly think that they are right

A voyage of this magnitude
Well what did we suppose?

Our lifetimes will be spent aboard
This is the home we chose

And so we keep on sailing
Though we are ill-at-ease
Because each and every one of us
Dreamed of these frontier seas

And gazing from the bow alone
At a faraway dust cloud
It looks as though a shore of light
But I know what it shrouds

The gaping maw of a chilling monster
The cosmonaut's greatest dread
That lurks in the center of our galaxy
Still hundreds of years ahead

And will not be for my eyes
But for those that will come after
And wrapped in my thoughts I turn
    for the bridge
Ignoring the crew's raucous laughter

For mine is only the first step
The *start* of a fantastic tale
That the bards and poets onboard
Will do their best to regale

And they'll tell of a stout hearted captain
That left well-known shores for a dream
And for something that faraway beckoned
Something he saw in a gleam

Something that with his last breath
And eyes growing steadily dimmer
The man in the Crow's nest shouted

Was getting brighter and rapidly nearer

And the poets will tell that in a sudden
    bright flash
The crew standing near were made blind
But the bards will assure that their sight
    was returned
But the good captain no one could find

This I have seen in the future
Written plain in the print of the stars
And outwardly I go about duties
While inwardly counting the hours

'Til the light I have seen approaches
And brightens the cheerless gloom
That has settled in the depths of my heart
And has shut out the light like a tomb

In the meantime I pace 'neath the rigging
The myriad network of ropes
Stopping in habit every now and then
To make a quick sweep with the scope

And the crew often glance at each other
"Ah captain you are watchful as always.
But why more so of late?"
Rather answer, I smile at their praise

I know that they blame paranoia on age
I have lived nearly a century now
Much longer than I expected
And in excellent health somehow

And my vision is clearer than ever
My senses are heightened and sharp
An old starfarer I may be

But a young starfarer at heart

For an excitement is building within me
And threatens to not be contained
A feeling like heat in my body-
A warmth spreading through my veins

A summons - o soul, do you hear it?
In the distance, a call to rise?
Awaken from feverish slumber
Something has been seen in the sky

"A light, Captain, getting closer.
I don't know whether comet or craft.
Captain? Sir, did you hear me?
There is nothing like it on our maps."

"Nor would there be." I quietly say
To the man bemused and abashed
"It is a transport vessel."
And with that is a sudden bright flash

Am I made the same as you?
Katherine Relf-Canas

The world's expert on synthetic lifeforms tells us "I used to live with a guy who loved paper books."

Aren't we all buried in fictional works, dreamwalking life, from beginning to end?

Picard's retirement, à la Montaigne, writing history people want to forget, in mournful decline, so it looks.

Data, now dead, sacrificed for the love of his friend.

It's not just Data the galaxy mourns; millions of Romulans lay dead, their avengers on Earth.

Spies of the round-ear Destroyer of Romulan myth, seek a sentient android of flesh and blood who harbors a secret self, occulted, a seed, in silence.

Synth life is outlawed; you see, these illicit daughters host Data's neurons from birth.

Their myth--if it's true--preview organics meeting insufferable violence!

Our little robot girls, fed on simulations, bump into their feigned humanity by examining their past.

Picard, ever the humanist, casts love wide; since to love another, an Other, is that not love, too?

Or is this the boundary of love in a galaxy so vast?

Don't you wonder--am I really me or am I made the same as you?

We all have fictions inside us, too, hidden beneath our existential crises; no, we don't own our own fabrication, so are we someone else's hologram?

Why invent an afterlife, a Heaven, why fear Hell, when on Earth we are also damned?

Author's Note: This poem is loosely based on the Shakespearean Sonnet. I retain the 14 lines and the rhyming scheme as well as the three quatrains and a couplet, but the poem is also, as you will see, in free verse, rather than the form's expected iambic pentameter. This results in a hybrid of two opposing structures, like Saturn squaring Neptune, or like the flesh and blood synths it was inspired by. I wrote it after I made a family challenge to my sister, teenager and husband to 'write a sonnet about Star Trek.' Sheltering project 101. I don't know where the impulse came as it had been a long time since I had watched any of the series or its spinoffs. I had homework to do; before I began this poem I watched the CBS series, Picard. I love how the first season ended with the Shakespearean quote from The Tempest: 'We are such stuff as dreams are made on.' I am now watching the Next Generation with plans to continue this series by going back to the beginning.

Not of Byzantium
Eric Robert Nolan

Awakening at one AM after dreaming
not of Byzantium,
not of Babylon, but better —
Not Shangri-La, but shaded limb —
The pine I climbed when I was nine.

No Acropolis, only
fallow farm and rising sun.
Across, a distant treeline
ascends to render Athens'
Parthenon prosaic.

Exceeding empires, exceeding
even Elysium, is
This slumber's ordinary boyhood field.

The Guy Belleranti Page

princess kisses frog
frog shapeshifts into a prince
princess turns to toad

her welcoming hug
can a loving kiss be next?
she bares hungry fangs

historical first
alien's organ transplant
even plays music

alien daughter
gives adolescent eye rolls
parents roll them back

found myself at last
wishing others could also
I am lost in space

Featured Poet: Avra Margariti

Greetings From

It comes in the mail,
hand-delivered by a very excited mailman
who claims he hasn't seen a postcard in years.

Greetings from Alpha Scorpii,
the colors rendered vivid and retro,
the cosmos twinkling around the star you
    traveled to
just to get away from me.

It wasn't even two months ago
that we were celebrating our anniversary,
swimming with the whales--not the actual,
    endangered beasts,
but the neon holographic colossi whose forms
    illuminated
your smiling eyes, bright as stars, wide as planets.

Three years together,
one crappy fight,
the decision to take a break from each other,
and all I get now is this cheery extraterrestrial
    postcard.

I flip it over, revealing your familiar grape-scented
    glitter scrawl:
Alpha Scorpii is actually a binary star system.

It's lonely up here without you.

Attached to the back,
a spaceship ticket
to you.

Greta Is Bringing the Drugs

sparkling mineral dust
scraped from the cave walls
of sleeping giants

golden liquid courage
begged from the land
of the leprechauns

solid soil clumps
of the Centauri variety
still echoing with hoof-falls

toxic-colored vapors
a jungle contained inside
a single Venus flytrap

Greta is bringing the drugs
soon we'll be drifting
celestial bodies falling into orbit
the Earth a bad taste in our mouths
left by some distant butterfly dream

if we're lucky.

The Birds and the Beasts

She was the gardener, plucking the hummingbirds
from the tigers' hungry maws, slicing
her arm to drizzle crimson dollops in the birdbath,
making homes for the small rabbits and rodents
among the folds of her skirt,
dropping bits of flesh like breadcrumbs in the
scintillating ponds
for the koi fish and geese to peck at.
She stood sentry in the sheep pen, next to the goats
bleating gently
and fluttering their eyelashes against her legs
the way the moths caressed her cheeks with their
wings.
When the wolves came, she let them take a few
sheep
baptized in her tears.
She knew that things had to die for other things to
grow.
And when the beasts' spirits took to the sky,
the tongue-pink worms and shiny beetles came out
to play,
feasting on the mountainous bodies.
At night, she hugged all the creatures to her bosom
--all her birds and her beasts,
all the rare flowers she had grown from nothing,
if her own vast being could be called nothing--
and slept in her Garden of Aether.

Nesting

It's nice being out in the open
Where they can stretch their molecules
And shake the static off their incorporeal limbs.
Spooking people
Making windowpanes rattle
Lightbulbs flicker
Women spit under their shirt collar
And men stomp their superstitious feet,
Now that's some good fun.
But at the end of the day, even naughty ghosts
Require rest.
They fold themselves up tight
And tuck into one another
Like spectral Matryoshka dolls,
Like flies burrowing under flesh.
Goodnight, they whisper to one another.
Goodnight, and thanks for the haunting.

What the Oneiroi Said

The last hours of sleep are always dreamless,
so that's when you will find us resting
and seeing to one another's wounds.

When we catch wild animals for our dinner,
we make sure their souls see verdant meadows
as they slip away.

We don't take pleasure in giving people nightmares.
It is our own terrors that we cannot stop
from trickling into our followers' dreams
like wine from a badly sealed amphora.

Did you know plants dream, too?
We've never witnessed such greenness
as in those chlorophyll sweet-dreams.

Don't go, dear Dawn.
With you by our side, we fall asleep safe,
we fall asleep sound.

Avra Margariti is a queer Social Work undergrad from Greece. She enjoys storytelling in all its forms and writes about diverse identities and experiences. Her work has appeared or is forthcoming in *Vastarien, Asimov's, Liminality, Arsenika, Star*Line, Eye to the Telescope,* and other venues. You can find her on twitter @avramargariti.

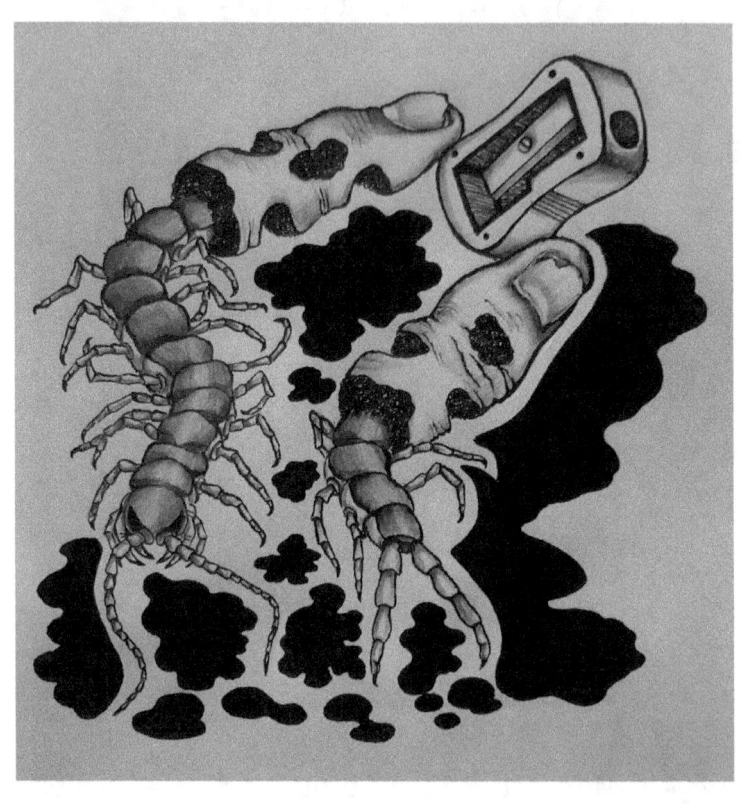

by Baishampayan Seal

Rumpled
Baishampayan Seal

Rumpled tulip petals
remnants of Adam's gift
before his plasma caster
decided to dig a hole
through his head

Rumpled tulip petals
Steve didn't even take
a look at them, or the bouquet
when it arrived this morning
The grey sticky note
didn't let him

Rumpled tulip petals
sitting together as pink flowers
had listened to the songs and kisses
of the first two androids ever
to pass the Turing test
rejuvenating the grass and the lake
unaware of the sin
the two bodies had committed
Their creators love to hate on
when someone parallels them
with the brain or the heart

Rumpled tulip petals
sitting together as black flowers
will never see
what happens to the lover
holding a suicide note
every cog, every circuit making
a death wish

Being chucked into the waste bin
doesn't let you anyway
They'll never know what
made Steve's better half
end his lithium-powered life
the burden of intelligence
or the burden of love

Zombies
Sean Ferrier-Watson

skin like wrinkled paper yellowed with age
they lumber down city streets
like old drunks looking for their last drink
or is it sleep
      restless
            moaning
                  desperate
                        for life
we see them and we don't
to see them is to join the plague
the endless shuffle down empty streets
the perpetual rise and fall of distant shadows
just out of reach
but we are the survivors
heroes one and all
with no time for the dead
in the night

The Color Green
Stephanie Smith

In a nowhere town, in a one-room shack,
sits a gray old man in a foil hat

He is ranting about UFO's
and how they're messing with
the television reception

They replaced his wife with a replica
who'll only cook peas for dinner
He doesn't like the color green
but eats them nonetheless

He claims they've done something with his head—
some sophisticated mind control
And the last few nights—as he lay in bed—
he's been transported someplace else:

to a land of computers, ten feet high
where beings speak in binary
beneath a green-warmed sky

Now, even though he can't understand
a single thing they say
their message is very clear

They send him home and back to bed
where he's left alone with his scattered thoughts
and a hankering left unfulfilled—
but that's why they're called dreams

In Feline Grace
Colleen Anderson

I love lying low in long grass
bathe in the sky's wild breath
hidden in shadows' light clasp
a fierce prowler watching
waiting, ready to spring
snare unwary with pointed fangs

My claws shift, shortening
from adamantine hardness, resolute
primal action into curled fingers
cringing, grasping for purchase
as boots stomp into ribs, my world
shifts and shakes, tenuous perch
swaying dropping me into
scalding words washing away
confidence and hope

I growl, hiss, swat away
crouch in the suffocating cave
of compliance, lick wounds
until the threat dissipates
alone in sunlight I groom, smoothing
hair, demeanor, determination

I will prove myself predator
free-spirited warrior, proud
when next a hand or foot aims
to damage, take me down
make me less, I will unleash my animal
self, use curved glinting claws
sharp as the blades in my drawer
and draw lines not to cross in red

Delivery
Lori R. Lopez

A symphony of moisture plays
its hollow taps and pings,
resounding in concordance,
a billion or so plucked strings.
Shotgun blasts of Thunder
crash and volley overhead,
as drops increase their tempo
and a stomach growls unfed —

Hunger swiftly answered
by a knell that chimes on cue.
Belly grumbling, I arise
to give the devil's due.
"Coming! Hold your horses!"
Eager fingers fumble locks,
then drag a heavy portal back,
revealing Chicken Pox . . .

Or some related ailment
resulting in red spots!
A pasty-visaged twerp behind
a rash of polkadots
toes the line of perspicacity,
clasping my To-Go.
He might've licked the Order,
snuck a taste for all I know.

Electric bolts, a blinking bulb
strobelight an anxious nerd.
I accept the sack and ditch it

with an understated word.
The only other sound the rain —
such music to my ears!
Delightedly I grab his shirt
and justify his fears.

The next ill-fated sap arrives,
hugging a plastic crate,
and blares "I'm from the Market!"
I greet my dinner date.
"Come inside. How nice to meet you!"
The grin a little broad.
My orbs the size of Tangerines,
imagining him clawed . . .

A pattern of bright hashmarks;
crisscrossed by scarlet stripes,
like the menu for a Werewolf
bearing tracks of wicked swipes
guaranteed to leave a scar,
but that isn't my affliction.
The handiwork I do
isn't based in lore and fiction.

"Just bring the items in."
"You sure that's all you need?"
He lugs four jugs of Bleach . . .
to scrub where morsels bleed.
"I'm on a special diet."
My edges become sharp
with talons and long cusps.
I roll him in a tarp.

This time no storm to muffle

a piercing shout or screech,
but I live out in the country
and lack neighbors it could reach.
My bane demanding human flesh
to satisfy Consumption
has kept me isolated,
running low on strength and gumption.

Relying on deliveries,
I had never learned to hunt.
There is no instinctive rage;
no desire to growl or grunt.
The world has its Infections
as I dwell in near-seclusion.
I only kill to eat.
That should clear up your confusion.

My invitation stands.
Should you wish to look around,
I'll restrain my urge to feed
in return for nothing found.
I claim the careless ones
who refuse to wear a mask
though they have no vaccination.
Stay away is all I ask.

Finally
Krista Canterbury Adams

Finally, we have come to the place
we wanted—
wooden tables stacked with fruit
and yeasty flat bread,
pitchers of spring water
with the green of mint,
the yellow of lemon
floating along the surface,
water said to be magic,
to be healing,
from an ancient sacred source.

We are in the side street,
in front of the spice shop
that sells cinnamon
and earthy nutmeg.
Here, in the sunlight,
for a while, we eat
of the magic of trees,
drink of the clear, fresh water.

Graveyard Waltz
Stephanie Smith

I float above the heads of the dead

Steal the jewels from their gem-encrusted crowns

I slip in and out of a hypnogogue

My lips form foreign words

From some musty forgotten text

I left at the foot of the bed

Incantations and ebon phrases

Black lace and roses in a vase I break

As the music stirs me from my reverie

The quiet night is shattered by the

Clicking of bones that try to grab me,

Grinding teeth that urge me to w*ake up!*

Who?

Lori R. Lopez is an offbeat hat-wearing speculative author, illustrator, poet, and songwriter residing in Southern California. Her prose and verse have been published in a number of anthologies and magazines. Book titles include The Dark Mister Snark, The Strange Tail of Oddzilla, Leery Lane, An Ill Wind Blows, The Fairy Fly, and Darkverse: The Shadow Hours (nominated for a 2018 Elgin Award). Two of her poems were nominated for Rhysling Awards. Lori co-owns Fairy Fly Entertainment with two talented sons. They've formed a band called The Fairyflies to release original music.

K.R.Lehman writes from a small town in Iowa. She is an avid aerospace, aviation, and science fiction enthusiast and is the author of the full-length poetry collection, "Moontouched, The Poetry of K.R.Lehman".

Stephanie Smith has been a contributor to the small press since she was a teenager in the nineties. Outside of writing she enjoys horror movies, Jeopardy!, and nature walks. She lives in Clarks Summit, Pennsylvania with her partner and recently adopted cat.

Meg Smith is a writer, journalist, dancer and events producer living in Lowell, Mass. In addition to previously appearing in Illumen, her poetry and fiction have appeared in The Horror Zine, Dream and Nightmares, Raven Cage, Dark Dossier, Sirens Call, and many more. She is the author of five poetry books. Her first short fiction collection, The Plague Confessor, is due out from Emu Books in fall 2020. She welcomes visits to megsmithwriter.com.

Russell Hemmell is a French-Italian transplant in Scotland, passionate about astrophysics, history, and Japanese manga. Recent poetry in Andromeda Spaceways Magazine, The Grievous Angel, Star*Line, and others. Find them online at their blog earthianhivemind.net and on Twitter @SPBianchini.

Krista Canterbury Adams has studied poetry at Ohio Dominican University and The Ohio State University in Columbus, Ohio. She has a six-poem series due to be printed this spring in AHF Magazine & has forthcoming work in His Dark Sire, Carmina, Collective Realms, CC&D magazines and BFS Horizons. She is a member of the SFPA.

Baishampayan Seal is based in Kolkata, India, where they are currently pursuing an MSc in Statistics. When not testing hypotheses or beating the keyboard for C++ or R coding, they enjoy writing short poetry and flash fiction. Their work has previously appeared in Aphelion Webzine and 365 Tomorrows, and is forthcoming in Scifaikuest, Night to Dawn and Bewildering Stories, among others. Find them on twitter @BaishampayanSe1.

William Landis: B.S. Agricultural Education Concentration Plant and Soil Science Class of 2012 North Carolina Agricultural and Technical State University

Colleen Anderson says: I am a Canada Council and BC Arts Council recipient in writing. My poetry placed in the Crucible, Wax Poetry, Rannu and Balticon contests. Some recent and forthcoming works are in *Eternal Haunted Summer, Space and Time, Quaranzine* and *Starline*. My fiction collection, *A Body of Work*, was published by Black Shuck Books, UK. I am currently shopping around two poetry collections.

www.ingramcontent.com/pod-product-compliance
Lightning Source LLC
LaVergne TN
LVHW011859060526
838200LV00054B/4419